Yohaku Junior

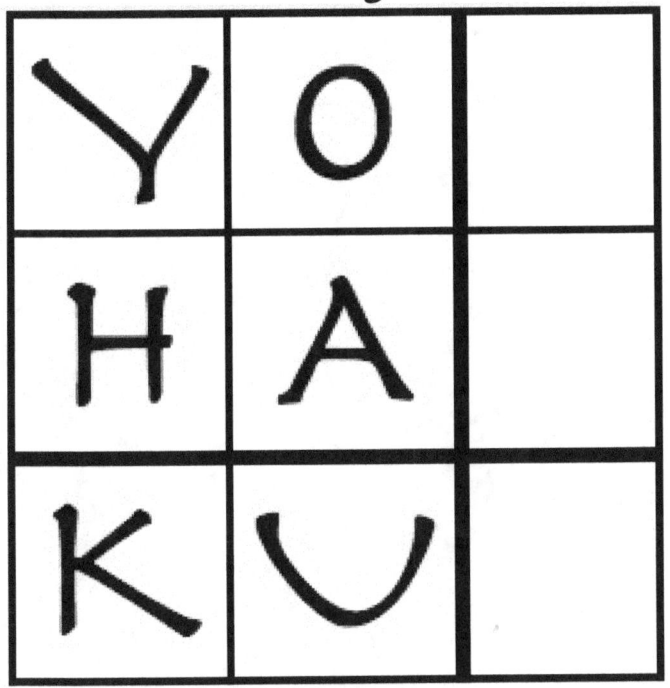

Book 3: More Additive and Multiplicative Puzzles

Mike Jacobs

For Màiri and Andrew,

and

for all the students, from all around the world, whom I
have had the pleasure of working with over the years. You
have made teaching feel like a lifetime in Paradise.

CONTENTS

Acknowledgments   i

1   What is a yohaku?   p. 1

2   Solving a yohaku   p. 2

3   Puzzles   p. 4

4   Solutions   p. 65

ACKNOWLEDGMENTS

Thanks must go out to all those who have encouraged me to write this book including my wife, Jackie, and my wonderful children Màiri and Andrew without whom…

I must also acknowledge all the wonderful maths colleagues that I have had the privilege of working with over the years in Yorkshire and in Ontario: you know who you are!

Finally, thank you to all the kind people who follow my Twitter account (@YohakuPuzzle) or use my website (www.yohaku.ca) for all their kind words of support and encouragement over the past few years.

1 WHAT IS YOHAKU JUNIOR?

Yohaku Junior™ is a new type of number puzzle that will test your number sense and problem solving skills. It is based on yohaku puzzles but is aimed at primary school-aged students. The puzzles in this book focus on additive and multiplicative thinking and will help students become more efficient with their recall of maths facts. Your task is to fill in the empty cells such that they give the sum shown in each row and column.

Each yohaku is lovingly created by hand!

2 SOLVING A YOHAKU

To solve a yohaku, you must fill in the blank space so that the cells give the sum or the product shown in each row and column.

In yohaku junior, one or more of the numbers are already given. For example, in the yohaku shown, the two empty cells in the top row must add to give 7.

3		7
		6
5	8	+

Since one of these cells is three, the other must be 4:

3	4	7
		6
5	8	+

Now we can use this to work out the other cells: the top two left column cells must add to give 5. Since one of these is 3, the other must be 2:

3	4	7
2		6
5	8	+

Finally the top two cells in the middle column must add to give 8. Since one of these is 4, the other must be 4:

3	4	7
2	4	6
5	8	+

We can check our answer by seeing that in the middle row, 2 plus 4 equals 6.

Some yohakus may have more than one possible solution. Many will involve a fair amount of trial and error.

The puzzles in this book are arranged from the easiest to the hardest. Solutions are provided. Happy solving!

3 PUZZLES

1

20		43
		46
44	45	+

2

30		55
		76
71	60	+

3

	40	80
		49
67	62	+

4

	32	52
		79
69	62	+

5

		54
32		64
65	53	+

6

		64
41		84
74	74	+

7

		80
	24	50
70	60	+

8

		67
	37	70
76	61	+

9

24		48
		90
72	66	+

10

31		70
		81
80	71	+

For puzzles 11 to 20, put the four given numbers into the blank cells to get the totals shown in each row and column.

11

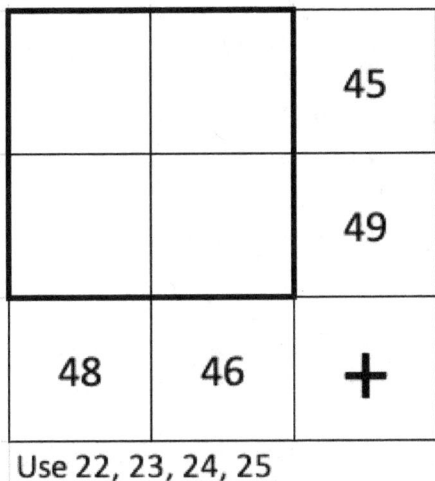

Use 22, 23, 24, 25

12

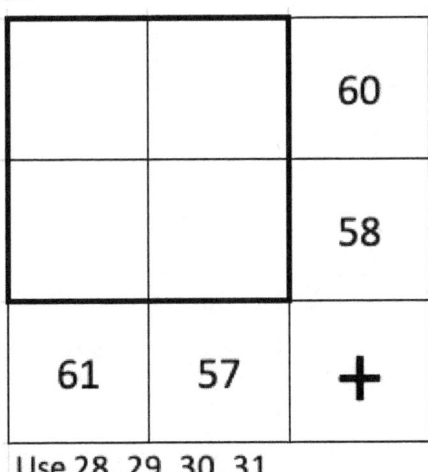

Use 28, 29, 30, 31

13

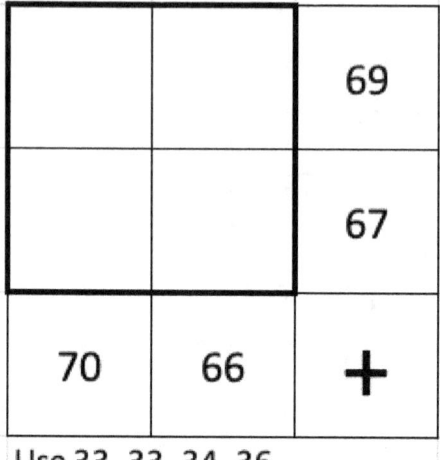

		69
		67
70	66	+

Use 33, 33, 34, 36

14

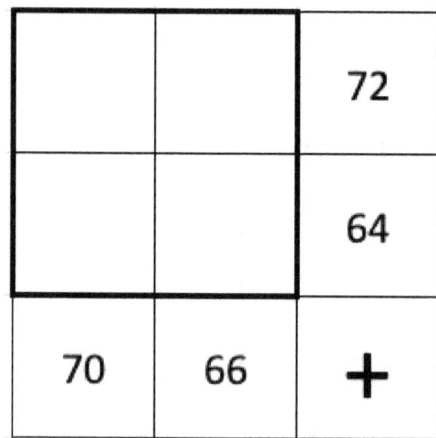

		72
		64
70	66	+

Use 31, 33, 35, 37

15

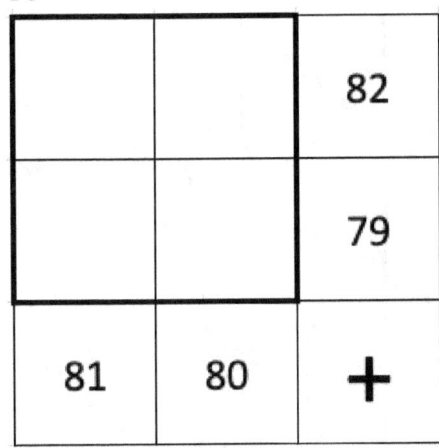

		68
		73
71	70	+

Use 32, 35, 36, 38

16

		82
		79
81	80	+

Use 37, 39, 42, 43

17

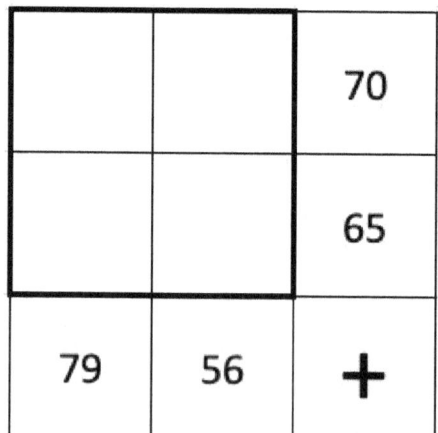

		80
		87
90	77	**+**

Use 36, 41, 44, 46

18

		70
		65
79	56	**+**

Use 23, 33, 37, 42

19

		78
		70
80	68	＋

Use 24, 34, 44, 46

20

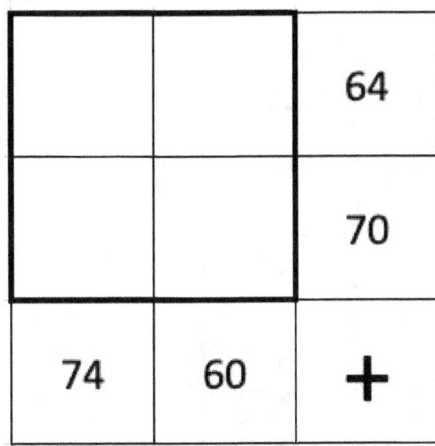

		64
		70
74	60	＋

Use 23, 27, 37, 47

21

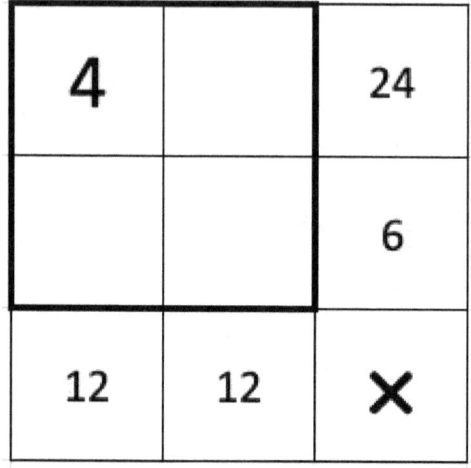

4		24
		6
12	12	✗

22

5		25
		24
20	30	✗

23

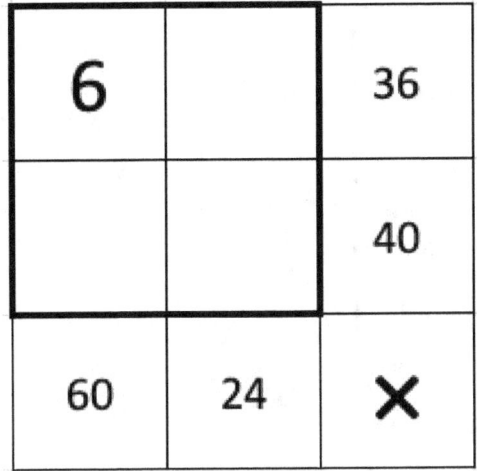

24

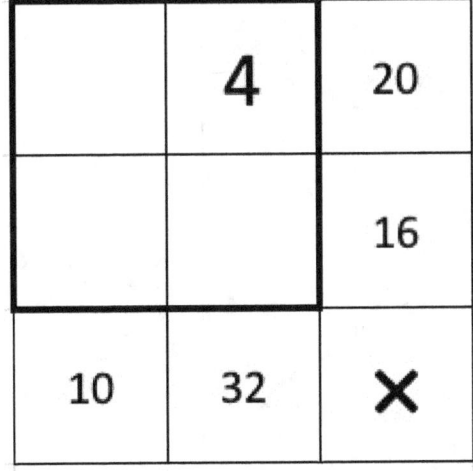

25

	5	40
		21
24	35	✕

26

	6	24
		21
28	18	✕

27

		24
2		18
16	27	✗

28

		35
4		32
28	40	✗

29

		24
	5	15
9	40	✕

30

		16
	6	42
28	24	✕

For puzzles 31 to 40, put the four given numbers into the blank cells to get the products shown in each row and column.

31

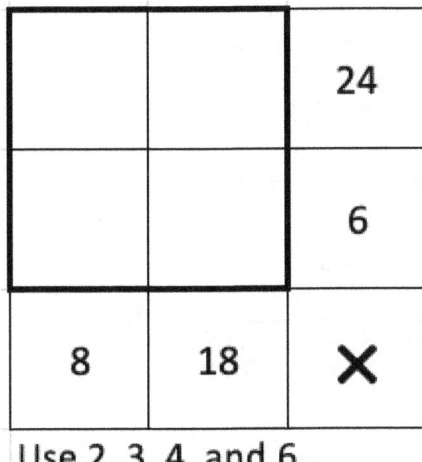

Use 2, 3, 4, and 6.

32

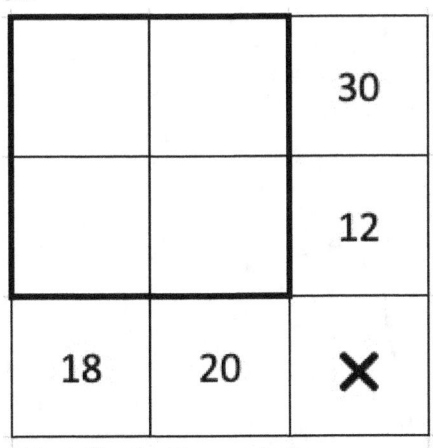

Use 3, 4, 5, and 6.

33

		24
		20
16	30	✗

Use 4, 4, 5, and 6.

34

		30
		18
15	36	✗

Use 3, 5, 6, and 6.

35

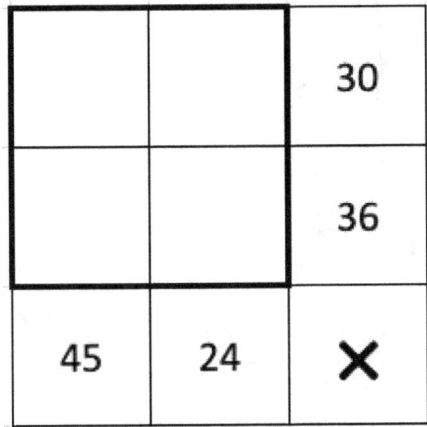

		21
		40
24	35	✗

Use 3, 5, 7, and 8.

36

		30
		36
45	24	✗

Use 4, 5, 6, and 9.

37

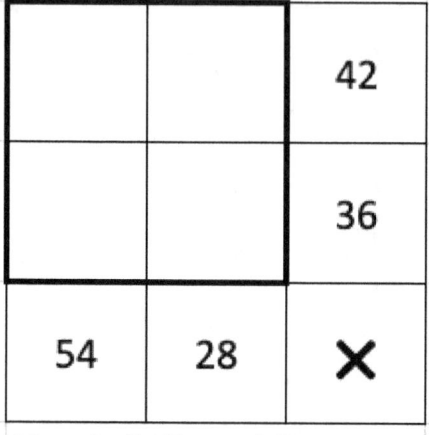

		42
		36
54	28	✗

Use 4, 6, 7, and 9.

38

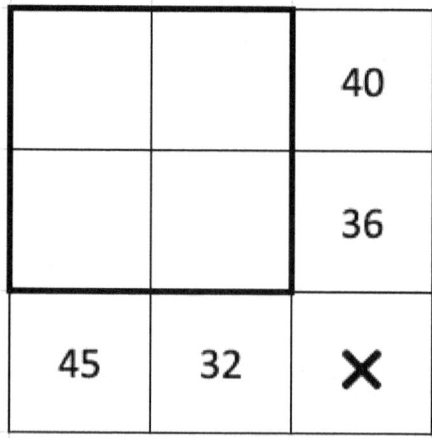

		40
		36
45	32	✗

Use 4, 5, 8, and 9.

39

		60
		40
30	80	✕

Use 5, 6, 8, and 10.

40

		30
		64
40	48	✕

Use 5, 6, 8, and 8.

41

52		83
		87
95	75	**+**

42

54		96
		90
99	87	**+**

43

58		91
		89
100	80	+

44

	62	96
		63
69	90	+

45

	65	107
		80
87	100	+

46

	68	92
		79
70	101	+

47

		63
75		90
100	53	**+**

48

		61
76		100
91	70	**+**

49

		74
	48	85
75	84	**+**

50

		73
	59	102
72	103	**+**

For puzzles 51 to 60, put the four given numbers into the blank cells to get the totals shown in each row and column.

51

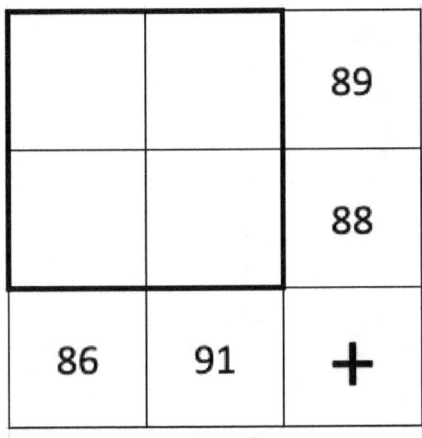

		89
		88
86	91	**+**

Use 42, 44, 45, 46

52

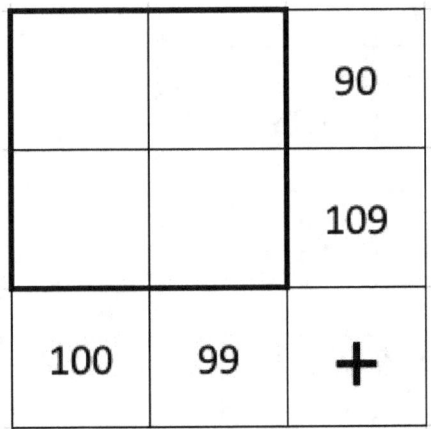

		90
		109
100	99	**+**

Use 38, 47, 52 and 62.

53

		111
		110
101	120	+

Use 46, 55, 55, and 65.

54

		104
		105
98	111	+

Use 42, 48, 56, and 63.

55

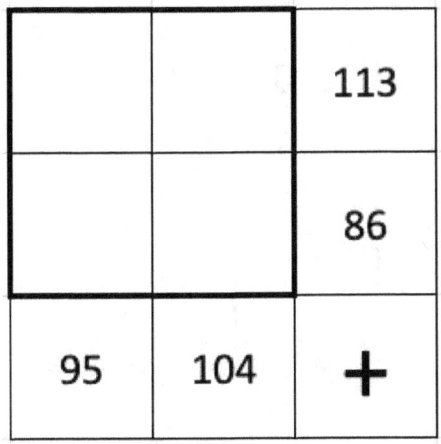

		112
		104
114	102	+

Use 46, 48, 56, and 66.

56

		113
		86
95	104	+

Use 23, 41, 63, and 72.

57

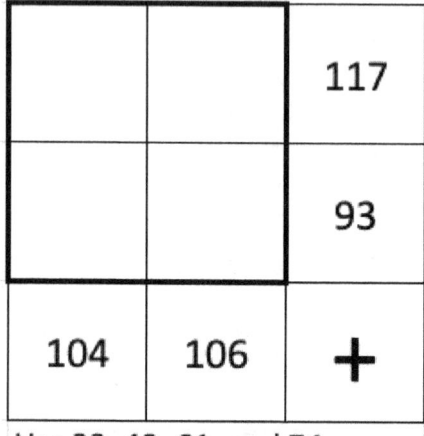

		117
		93
104	106	➕

Use 32, 43, 61, and 74.

58

		110
		100
90	120	➕

Use 24, 44, 66, and 76.

59

		101
		105
117	89	✛

Use 26, 42, 63, and 75.

60

		126
		94
113	107	✛

Use 29, 42, 65 and 84.

61

7		28
		30
35	24	✕

62

7		42
		24
21	48	✕

63

8		40
		28
32	35	✗

64

	8	48
		42
36	56	✗

65

	7	63
		21
27	49	✕

66

	9	45
		32
40	36	✕

67

		30
8		64
48	40	✗

68

		42
7		56
49	48	✗

69

		32
	9	54
48	36	✗

70

		49
	9	72
56	63	✗

For puzzles 71 to 80, put the four given numbers into the blank cells to get the products shown in each row and column.

71

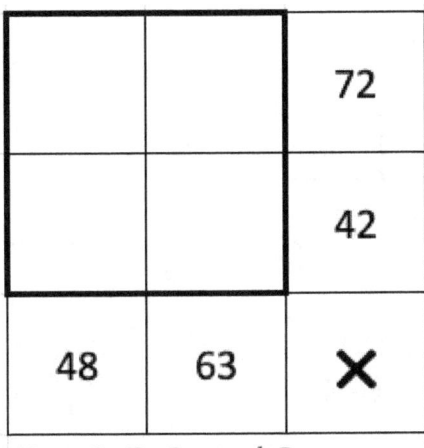

Use 6, 7, 8, and 9.

72

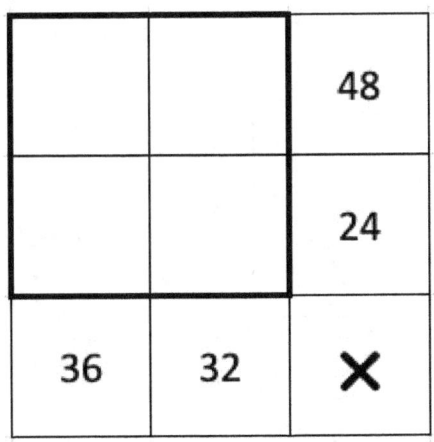

Use 3, 4, 8, and 12.

Mike Jacobs

73

		55
		54
45	66	✕

Use 5, 6, 9, and 11.

74

		48
		54
36	72	✕

Use 4, 6, 9, and 12.

40

75

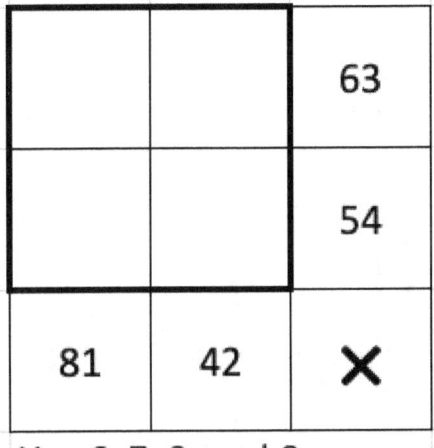

		63
		54
81	42	✗

Use 6, 7, 9, and 9.

76

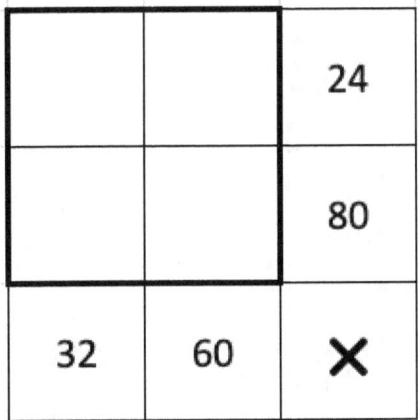

		24
		80
32	60	✗

Use 3, 4, 8, and 20.

77

		14
		48
28	24	✕

Use 2, 4, 7, and 12.

78

		54
		77
63	66	✕

Use 6, 7, 9, and 11.

79

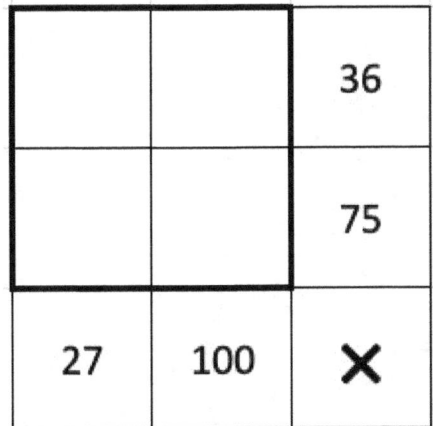

		60
		66
72	55	✗

Use 5, 6, 11, and 12.

80

		36
		75
27	100	✗

Use 3, 4, 9, and 25.

81

10	9		24
	8	4	18
			30
27	29	16	+

82

		8	32
	3	17	40
5			40
37	35	40	+

83

			40
	6	2	24
4	13		31
35	30	30	+

84

		7	24
	11		33
15	4		39
35	24	37	+

85

		18	51
	10		36
14	13		39
45	40	41	+

86

21	20		48
9	30		43
			42
46	55	32	+

87

	20	30	61
	24	23	60
			47
38	60	70	+

88

	50	20	101
	40	30	100
			102
101	101	101	+

89

	35	40	100
		46	70
21			70
70	80	90	**+**

90

	19	12	45
		17	45
18			45
45	45	45	**+**

Notice that this is a magic square: each row, column and diagonal add up to the same number!

91

1	2		8
2	5		100
			36
6	30	160	✖

92

3	1		12
2	2		16
			42
18	14	32	✖

93

			80
3	4		36
3	3		9
45	24	24	✗

94

		6	54
	7	2	14
8			32
24	21	48	✗

95

		2	40
	4	7	28
	2		54
36	40	42	✕

96

			24
2	5		70
	3	3	72
48	60	42	✕

97

3	2		48
3			96
		2	18
81	16	64	✕

98

			81
2		6	36
2		5	50
36	135	30	✕

99

2	3		54
5		8	80
			49
70	42	72	✕

100

	3	2	66
5		4	40
			81
165	18	72	✕

101

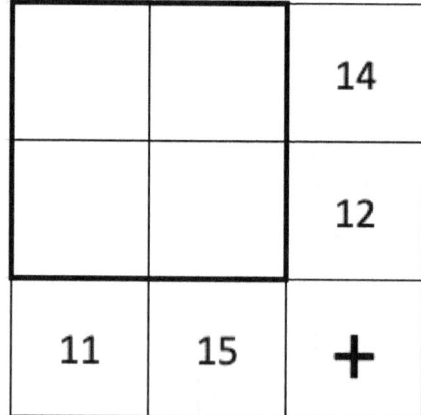

		9
		5
6	8	+

Use 4 consecutive numbers.

102

		14
		12
11	15	+

Use 4 consecutive numbers.

103

		12
		14
15	11	+

Use 4 consecutive numbers.

104

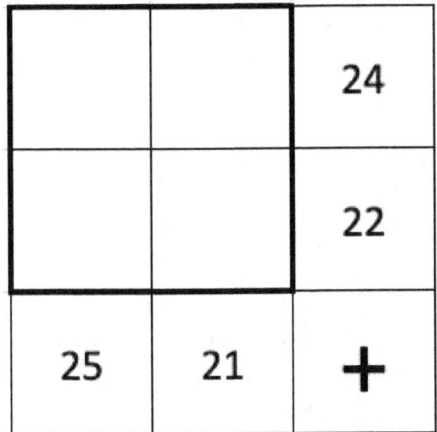

		24
		22
25	21	+

Use 4 consecutive numbers.

105

		23
		27
26	24	+

Use 4 consecutive numbers.

106

		28
		30
27	31	+

Use 4 consecutive numbers.

107

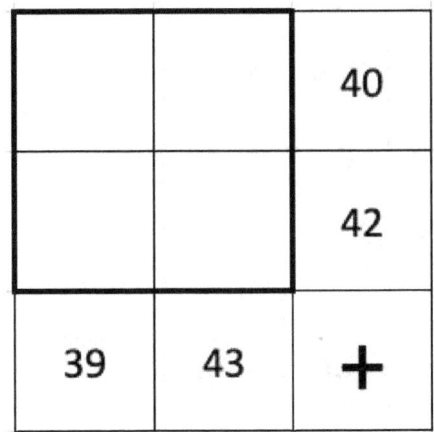

		44
		42
45	41	+

Use 4 consecutive numbers.

108

		40
		42
39	43	+

Use 4 consecutive numbers.

109

		15
		14
35	6	✕

Use 4 different numbers.

110

		8
		15
12	10	✕

Use 4 different numbers.

111

		9
		24
27	8	✗

Use 4 different numbers.

112

		30
		7
10	21	✗

Use 4 different numbers.

113

		20
		18
8	45	✕

Use 4 different numbers.

114

		18
		28
21	24	✕

Use 4 different numbers.

115

		24
		36
32	27	✕

Use 4 different numbers.

116

		36
		35
20	63	✕

Use 4 different numbers.

Mike Jacobs

117

			6
			17
			22
10	19	16	**+**

Use 9 consecutive numbers.

118

			42
			50
			6
2	70	90	**✕**

Use 1, 1, 2, 2, 3, 5, 5, 6, and 7.

119

Can you find three different solutions for this yohaku?

		36
		32
16	72	✗

Use 4 different whole numbers.

		36
		32
16	72	✗

Use 4 different whole numbers.

		36
		32
16	72	✗

Use 4 different whole numbers.

120 Can you find six different solutions for this yohaku?

		24
		36
12	72	✕

Use 4 different whole numbers.

		24
		36
12	72	✕

Use 4 different whole numbers.

		24
		36
12	72	✕

Use 4 different whole numbers.

		24
		36
12	72	✕

Use 4 different whole numbers.

		24
		36
12	72	✕

Use 4 different whole numbers.

		24
		36
12	72	✕

Use 4 different whole numbers.

4 SOLUTIONS

1

20	23	43
24	22	46
44	45	+

2

30	25	55
41	35	76
71	60	+

3

40	40	80
27	22	49
67	62	+

4

20	32	52
49	30	79
69	62	+

5

33	21	54
32	32	64
65	53	+

6

33	31	64
41	43	84
74	74	+

7

44	36	80
26	24	50
70	60	+

8

43	24	67
33	37	70
76	61	+

9

24	24	48
48	42	90
72	66	+

10

31	39	70
49	32	81
80	71	+

11

23	22	45
25	24	49
48	46	+

12

31	29	60
30	28	58
61	57	+

13

36	33	69
34	33	67
70	66	+

14

37	35	72
33	31	64
70	66	+

15

36	32	68
35	38	73
71	70	+

16

39	43	82
42	37	79
81	80	+

17

44	36	80
46	41	87
90	77	+

18

37	33	70
42	23	65
79	56	+

19

34	44	78
46	24	70
80	68	+

20

27	37	64
47	23	70
74	60	+

21

4	6	24
3	2	6
12	12	×

22

5	5	25
4	6	24
20	30	×

23

6	6	36
10	4	40
60	24	×

24

5	4	20
2	8	16
10	32	×

25

8	5	40
3	7	21
24	35	×

26

4	6	24
7	3	21
28	18	×

27

8	3	24
2	9	18
16	27	×

28

7	5	35
4	8	32
28	40	✕

29

3	8	24
3	5	15
9	40	✕

30

4	4	16
7	6	42
28	24	✕

31

4	6	24
2	3	6
8	18	✕

32

6	5	30
3	4	12
18	20	✕

33

4	6	24
4	5	20
16	30	✕

34

5	6	30
3	6	18
15	36	✕

35

3	7	21
8	5	40
24	35	✕

36

5	6	30
9	4	36
45	24	✕

37

6	7	42
9	4	36
54	28	✕

38

5	8	40
9	4	36
45	32	✕

39

6	10	60
5	8	40
30	80	✕

40

5	6	30
8	8	64
40	48	✕

41

52	31	83
43	44	87
95	75	✚

42

54	42	96
45	45	90
99	87	✚

43

58	33	91
42	47	89
100	80	+

44

34	62	96
35	28	63
69	90	+

45

42	65	107
45	35	80
87	100	+

46

24	68	92
46	33	79
70	101	+

47

25	38	63
75	15	90
100	53	+

48

15	46	61
76	24	100
91	70	+

49

38	36	74
37	48	85
75	84	+

50

29	44	73
43	59	102
72	103	+

51

44	45	89
42	46	88
86	91	+

52

38	52	90
62	47	109
100	99	+

53

46	65	111
55	55	110
101	120	+

54

56	48	104
42	63	105
98	111	+

55

66	46	112
48	56	104
114	102	+

56

72	41	113
23	63	86
95	104	+

57

43	74	117
61	32	93
104	106	+

58

66	44	110
24	76	100
90	120	+

59

75	26	101
42	63	105
117	89	+

60

84	42	126
29	65	94
113	107	+

61

7	4	28
5	6	30
35	24	×

62

7	6	42
3	8	24
21	48	×

63

8	5	40
4	7	28
32	35	×

64

6	8	48
6	7	42
36	56	×

65

9	7	63
3	7	21
27	49	×

66

5	9	45
8	4	32
40	36	×

67

6	5	30
8	8	64
48	40	×

68

7	6	42
7	8	56
49	48	×

69

8	4	32
6	9	54
48	36	×

70

7	7	49
8	9	72
56	63	×

71

8	9	72
6	7	42
48	63	×

72

12	4	48
3	8	24
36	32	×

73

5	11	55
9	6	54
45	66	✕

74

4	12	48
9	6	54
36	72	✕

75

9	7	63
9	6	54
81	42	✕

76

8	3	24
4	20	80
32	60	✕

77

7	2	14
4	12	48
28	24	✕

78

9	6	54
7	11	77
63	66	✕

79

12	5	60
6	11	66
72	55	✕

80

9	4	36
3	25	75
27	100	✕

81

10	9	5	24
6	8	4	18
11	12	7	30
27	29	16	+

82

12	12	8	32
20	3	17	40
5	20	15	40
37	35	40	+

83

15	11	14	40
16	6	2	24
4	13	14	31
35	30	30	+

84

8	9	7	24
12	11	10	33
15	4	20	39
35	24	37	+

85

16	17	18	51
15	10	11	36
14	13	12	39
45	40	41	+

86

21	20	7	48
9	30	4	43
16	5	21	42
46	55	32	+

87

11	20	30	61
13	24	23	60
14	16	17	47
38	60	70	+

88

31	50	20	101
30	40	30	100
40	11	51	102
101	101	101	+

89

25	35	40	100
24	0	46	70
21	45	4	70
70	80	90	+

90

14	19	12	45
13	15	17	45
18	11	16	45
45	45	45	+

91

1	2	4	8
2	5	10	100
3	3	4	36
6	30	160	×

92

3	1	4	12
2	2	4	16
3	7	2	42
18	14	32	×

93

5	2	8	80
3	4	3	36
3	3	1	9
45	24	24	×

94

3	3	6	54
1	7	2	14
8	1	4	32
24	21	48	×

95

4	5	2	40
1	4	7	28
9	2	3	54
36	40	42	×

96

3	4	2	24
2	5	7	70
8	3	3	72
48	60	42	×

97

3	2	8	48
3	8	4	96
9	1	2	18
81	16	64	×

98

9	9	1	81
2	3	6	36
2	5	5	50
36	135	30	×

99

2	3	9	54
5	2	8	80
7	7	1	49
70	42	72	×

100

11	3	2	66
5	2	4	40
3	3	9	81
165	18	72	×

101

4	5	9
2	3	5
6	8	+

102

6	8	14
5	7	12
11	15	+

103

7	5	12
8	6	14
15	11	+

104

13	11	24
12	10	22
25	21	+

105

12	11	23
14	13	27
26	24	+

106

13	15	28
14	16	30
27	31	+

107

23	21	44
22	20	42
45	41	+

108

19	21	40
20	22	42
39	43	+

109

5	3	15
7	2	14
35	6	×

110

4	2	8
3	5	15
12	10	×

111

9	1	9
3	8	24
27	8	×

112

10	3	30
1	7	7
10	21	×

113

4	5	20
2	9	18
8	45	×

114

3	6	18
7	4	28
21	24	×

115

8	3	24
4	9	36
32	27	×

116

4	9	36
5	7	35
20	63	×

117

1	3	2	6
4	7	6	17
5	9	8	22
10	19	16	+

118

1	7	6	42
2	5	5	50
1	2	3	6
2	70	90	✕

119 (1)

1	36	36
16	2	32
16	72	✕

119 (2)

2	18	36
8	4	32
16	72	✕

119 (3)

4	9	36
4	8	32
16	72	✕

120 (1)

1	24	24
12	3	36
12	72	✕

120 (2)

2	12	24
6	6	36
12	72	✕

120 (3)

3	8	24
4	9	36
12	72	✕

120 (4)

4	6	24
3	12	36
12	72	✕

120 (5)

6	4	24
2	18	36
12	72	✕

120 (6)

12	2	24
1	36	36
12	72	✕

ABOUT THE AUTHOR

Mike Jacobs is a Mathematics teacher living in Ontario, Canada. Originally from Yorkshire in the U.K., he has been teaching for 29 years. He first fell in love with mathematics after watching Disney's Donald in Mathemagic Land. He created yohaku puzzles as a means of encouraging problem solving as well as practising and improving number sense. His favourite prime number is 23 and he is convinced that triangle numbers are much more cooler than square numbers.